AF799936

Ditch The Store series

From Brooder to Plate

The Step-by-Step Guide to Raising Chickens for Meat

Published by CJ Steedman

Copyright © 2025 All rights reserved.
No part of this publication may be copied or reproduced in any format, by any means, electronic or otherwise, without prior consent from the copyright owner and publisher of this book.

ISBN: 978-1-7638806-0-3

DITCH THE STORE

From Brooder to Plate

Dedication

Without my amazing family

none of this would be possible.

To the Handy Helper, MJ and OJ,

Thank you

Table of Contents

Introduction

G'day and welcome to the first book in the Ditch the Store series. Firstly, I'd like to say congratulations on making the decision to Ditch the Store and raise your own meat. The choice to go into homesteading is not a decision to be entered into lightly.

Thank you for choosing me to help you on that journey. I'm especially excited to be the one to help you in this process.

A little bit about me so that you know why you can trust me to help you on this journey.

My name is CJ Steedman, and I live in Yass River NSW, Australia, on 120 acres with my family. I grew up on a farm with my parents; however, I left in my teens to find work. As an adult, I decided that farming was too ingrained in my blood, and I wanted to go back to the land.
Luckily, even though I was a single mum at the time, I met a wonderful man who has never said no to any of my crazy schemes or dreams. I refer to him often as the Handy Helper, and without his assistance, I would have never been able to do the things that I've done.

I have pasture-raised chickens for meat and eggs. I also raise goats for meat, fibre and milk. The farm is expanding to include beef and dairy cattle, and I'm also a beekeeper. Deep down, I consider myself to be a true homesteader, as I try as hard as I can not to purchase anything from the supermarkets.

Obviously, in this modern day and age, that's a really hard thing to do, but I do believe that we can get to a point where we are only buying the bare minimum from a supermarket or store. Toilet paper, for example, is obviously one thing that we will never be able to replace, although I have considered going to washable wipes; that's really stretching the friendship with my family. Pretty much everything else in my house could be grown, raised, or made within my home

This series of books is designed to help you do what I'm doing, which is breaking away from the horrible cycle of being stuck at the supermarket and instead being as self-sufficient as possible so that you too can Ditch the Store.

I hope that this series will give you the information you need to make this journey, and I would love to hear your story. You can reach me at Mojo Homestead on all the social sites, and I encourage you to link Mojo Homestead to pictures of your achievements. I will be your biggest cheerleader.

If you like podcasts, you can find me on Spotify by searching for "Not the Farmers Wife, with a new episode each week.

Otherwise, please enjoy the book, and if you get value from it, please consider reviewing it so others will find it too.

CJ

Chapter 1

Introduction to Meat Chickens

Understanding Meat Chicken Breeds

So, you've decided to raise chickens for meat. Welcome to the world of plump, delicious birds that can transform your homestead and your kitchen. But wait—before you run out and grab every chicken you can find, let's take a minute to understand what makes meat chickens tick and which breeds are best suited for your needs. It's more than just picking up a few random chickens at the feed store and calling it a day.

You've got decisions to make, and we're going to break them down so you get it right from the start. How does that saying go,

Start how you intend to finish.

Most importantly, I have made a number of mistakes in raising meat birds—mistakes I don't want you to make. This book and corresponding online course have been put together to help you have an easier time raising your first meat birds than I did.

The World of Meat Chickens: What's the Deal with These Plump Birds?

When most people think of chickens, they think of eggs. But meat chickens are a different breed of bird altogether—pun intended! These chickens are bred specifically for their meat production, not for egg-laying.

While you can raise dual-purpose chickens (the kind that lay eggs and will grow to a large, edible bird), meat-specific breeds are designed to grow fast and fatten up with as little effort as possible on your part.

Meat chickens are typically larger, with more muscle mass, and they reach their ideal size much quicker than your average backyard egg-laying hens. We're talking a full-grown bird in about 8 to 10 weeks, ready for your table. The best part? They've been bred for this. No surprises, no guesswork. These birds have been genetically designed to grow big, fast, and efficiently.

If you're going to raise chickens for meat, you need to think about what's going to work best for your homestead.

Do you want a chicken that's ready to eat in under 10 weeks?

Are you interested in efficiency?

Or do you want a bird that's going to take a little longer but might have a bit more personality (and flavour)?

Let's dig into the differences.

Meat Specific Breeds vs. Heritage Breeds: What's Best for Your Homestead?

Here's the deal—meat-specific breeds are all about efficiency. These chickens are bred for one reason: to put on weight fast and provide maximum meat in the shortest amount of time.

Figure 1 Heritage Breed Brahma versus Hybrid Cob

On the flip side, heritage breeds (the ones that are older and more traditional breeds) are a bit slower. They take longer to reach full size, but they might have more flavour or a more interesting story behind them.

So, which is right for your homestead? Let's look at a couple of the options.

Cornish Cross: The most common breed for meat chickens, and for good reason. These birds are what you see in the supermarket: plump, juicy, and ready for harvest in 8 to 10 weeks. The Cornish Cross grows at an astonishing rate, making it a top choice for homesteaders looking to fill their freezer quickly and efficiently.

The downside? These chickens are known for having a bit of a "sitting around" nature, so you'll need to make sure they have enough room to move around in their pen without sitting on top of each other. But if speed and efficiency are your main goals, these are the birds for you. Cornish Cross in Australia is often called a "cob."

Red Rangers: These guys are a good middle ground between the speedy Cornish Cross and the slow-growing heritage breeds. Taking 11–12 weeks to grow to full size, they are also much more weather-hardy than the Cornish Cross. They are often raised on pasture and are known for being a bit more active, which is a great thing if you want a bird that's going to enjoy a little freedom before it ends up on your dinner plate. Red Rangers can lay eggs, but they aren't considered to be dual purpose and would be unreliable in that role.

Freedom Rangers: Another excellent choice for pasture-raised birds. Freedom Rangers grow at a slower pace than Cornish Cross, taking 11–12 weeks to grow to full weight. They are much more active, giving them a better quality of life if you're into that sort of thing. They also have a reputation for having better flavour, which makes them a favourite among some homesteaders. Interestingly, Freedom Rangers will lay eggs and can breed.

Heritage Breeds: While these birds, like the Rhode Island Red or Plymouth Rock, are known for their egg-laying prowess, they can also be raised for meat. They take longer to grow, but many homesteaders appreciate the flavour and the slower pace of life that comes with raising them. If you're after something that's a bit more "natural" and want to embrace old-school homesteading vibes, heritage breeds might be up your alley.

Growth Rates: How Fast Do You Want Them?

Ah, the million-dollar question—how fast do you want your chickens to grow? Well, that depends on your goals. If you want a bird that's ready for harvest in a matter of weeks, then the Cornish Cross/Cob is your best friend. These birds will get big and fast and will be ready to eat in just about 2 months. This is perfect for those of you who want to get the job done quickly and efficiently.

But not everyone is in a rush. If you have the space and time to let your birds grow at a slower pace, then you might opt for something like the Red Ranger or the Freedom Ranger. These birds take longer to mature, but they tend to have better flavour and more "real" chicken texture than their super-fast-growing counterparts.

I have heard some people don't like the flavour of the slower growers. I think most family's palates have adapted to the quick-growing hybrid birds that are favourites for commercial growers. This adaptation means when we taste traditional meat, it can sometimes come across as "gamey" and not to everyone's tastes. This is not to put you off heritage breeds, as having tasted both, I know I could adjust very quickly to the heritage meat. It is something to consider in your choices, particularly if you have any fussy eaters.

It's also worth considering the time of year. If you're raising meat chickens in the warmer months, you might find that the faster-growing breeds like the Cornish Cross are better suited for the season. In the winter, you'll have a bit more time for slower-growing breeds to put on the weight, and you'll want a cold, hardy bird.

Setting Goals for Meat Production

Now that we've covered the breeds, let's talk about how to make this whole chicken-raising thing work for you. It's all well and good to have a flock of chickens, but how many should you actually raise? How do you plan for seasonal consumption? If you are raising chickens for eggs you'll have to wait for my next book as that will cover the topic of our egg laying flocks.

How Many Chickens Should You Raise for Meat?

You've got space, you've got the time, and now it's time to do the math. How many chickens do you need to raise to keep your family fed for the year?

On average, a full-grown meat chicken will yield about 1.3-1.8 kg (or 3 to 4 pounds) of usable meat, depending on the breed and how you process it. You can calculate by weight of meat consumed; however, I find it easier to say my family eats one or two whole birds per week.

Let's say you and your family eat one whole chicken once a week (if you're anything like my family, it's more often than that). That's 52 meals per year. If you're aiming to raise enough meat chickens to cover a full year's worth of chicken dinners, it is an easy calculation of 52 birds per year. By that scale, you'll need to raise about 52 chickens for the year. You can adjust the numbers based on your family's eating habits or if you want some extra birds for a few special meals or sharing with friends.

Chicken Consumption Table

Weekly Consumption (Whole Chickens)	Annual Consumption (Whole Chickens)	Over 5 batches
1	52	10
2	104	21
3	156	31
4	208	42
5	260	52
6	312	62
7	364	73
8	416	83
9	468	94
10	520	104

Planning for Seasonal Consumption:

When to Raise, Harvest, and Eat

Timing is everything when it comes to raising chickens for meat. You don't want to have all your chickens ready to go in the middle of winter when it's freezing outside and processing them becomes a huge hassle.

Planning for seasonal consumption is key.

For those of you living in colder climates, you'll want to raise your chickens so that they're ready to harvest in the late spring or early fall.

This way, you're not working in freezing temperatures, and your birds will have had plenty of time to enjoy the outdoors and eat the grass and bugs they love.

For warmer climates, it's a little easier—you can raise chickens year round, as long as you're providing them with shade and water during the hotter months.

Some homesteaders plan for a spring and fall harvest, allowing them to have fresh meat on hand when it's most convenient weather-wise.

Sample of 24 Week/ 5 Batch Schedule for raising meat birds

Week	Batch 1	Batch 2	Batch 3	Batch 4	Batch 5
WK 1	Brooder: Day-old chicks				
WK. 2	Brooder: Week Old Chicks				
WK. 3	Brooder: Prep to move out				
WK. 4	Chicken Tractor: Move to grass	Order next batch of day-old			
WK. 5	Chicken Tractor: Pasture raised	Brooder: Day-old chicks			
WK. 6	Chicken Tractor: Pasture raised	Brooder: Week Old Chicks			
WK. 7	Chicken Tractor: Pasture raised	Brooder: Prep to move out			
WK. 8	Prepare for processing	Chicken Tractor: Move to grass	Order next batch of day-old		
WK. 9		Chicken Tractor: Pasture raised	Brooder: Day-old chicks		
WK 10		Chicken Tractor: Pasture raised	Brooder: Week Old Chicks		
WK 11		Chicken Tractor: Pasture raised	Brooder: Prep to move out		
WK 12		Prepare for processing	Chicken Tractor: Move to grass	Order next batch of day-old	
WK 13			Chicken Tractor: Pasture raised	Brooder: Day-old chicks	
WK 14			Chicken Tractor: Pasture raised	Brooder: Week Old Chicks	
WK 15			Chicken Tractor: Pasture raised	Brooder: Prep to move out	
WK 16			Prepare for processing	Chicken Tractor: Move to grass	Order next batch of day-old
WK 17				Chicken Tractor: Pasture raised	Brooder: Day-old chicks
WK 18				Chicken Tractor: Pasture raised	Brooder: Week Old Chicks
WK 19				Chicken Tractor: Pasture raised	Brooder: Prep to move out
WK 20				Prepare for processing	Chicken Tractor: Move to grass
WK 21					Chicken Tractor: Pasture raised
WK 22					Chicken Tractor: Pasture raised
WK 23					Chicken Tractor: Pasture raised
WK 24					Prepare for processing

Is It All About Eggs? Nope, It's About Protein,

I know, I know, eggs are awesome. They're like little protein-packed nuggets of joy, and having fresh eggs on hand is one of the best parts of owning chickens. But when it comes to raising chickens for meat, eggs are a nice bonus, but let's be real—this is about protein.

Raising meat chickens is primarily about providing food for your family in a sustainable way. Whether you're looking to cut back on grocery bills, grow your own food, or just take control of your food supply, meat chickens give you that freedom.

They're an investment in self-sufficiency, and this will provide your family with an amazing source of protein for months to come. Plus, you'll know exactly where your food comes from and how it was raised.

So, let's recap: Raising meat chickens is about more than just filling your freezer. It's about becoming more self-sufficient and taking control of your food production.

It's about the satisfaction of knowing that your family is eating healthy, home-raised, homegrown food that you worked hard for. And trust me, nothing tastes better than a chicken you raised yourself.

Chapter 2

Starting with Day-Old Chicks

So, you're ready to dive into raising meat chickens. Maybe you've been thinking about it for a while, or maybe you just woke up one day and thought, "Yeah, I can totally do this."

Whatever the reason, you're about to embark on an exciting (sometimes stressful but often rewarding) journey. But before you get too far ahead of yourself, let's talk about where it all starts: those adorable, chirping, fluffy little day-old chicks.

They might look like they're just cute balls of fluff, but they're your ticket to a self-sufficient, home-raised, and—most importantly—delicious future.

I often joke with my students that,

Chickens are the gateway drug to homesteading

It's a slippery slope once to start, and before you know it, you'll have milking goats, bees, and a full vegetable garden.

In this chapter, we're going to break down everything you need to know about getting your chicks. From sourcing them to setting up a brooder that even your old farm-based grandma would approve of, I've got you covered.

So, let's get started—because these chicks are going to grow fast, and before you know it, you'll be chasing them around in their fancy new pasture homes.

All these principles apply for egg-laying chickens also, at least up until week 3. It is vital that you never put birds that have not developed their adult feathers outside overnight; cold and wet can kill them if they do not have their full feathers.

Sourcing Day-Old Chicks

Let's be real for a minute. You've decided to raise chickens for meat, but now comes the first hurdle: where do you get them?

Figure 2 Chicks less than 2 weeks

There's a lot of options out there, and it can get a little overwhelming. I want you to not stress about this decision; just weigh up the pros and cons and then do it!

If you find it didn't work as well as you thought or there were some problems with the choice you made, that's fine. Just make a different choice next time.

The fact that you are taking control of your food supply puts you way ahead of so many other families either way.

Where to Buy Chicks: Hatcheries vs. Local Farms (Spoiler Alert: You Want Local)

When it comes to sourcing day-old chicks, you've got two main options: hatcheries or local farms. It is an individual choice, and both have pros and cons.

Hatcheries are convenient—they've got a wide selection of chicks, they'll ship them right to your door, and you can pretty much order any breed your heart desires.

Something to keep in mind is that chicks that are shipped, especially from large hatcheries, are often stressed by the long journey. By the time they arrive, they've been cooped up (pun intended) in boxes for hours, maybe days, and that stress can take a toll on their health.

And if you're a beginner, it can be frustrating to have sickly chicks that are already fighting an uphill battle.

Local farms, on the other hand, offer chicks that are much more likely to be healthy and acclimated to your local environment. Plus, you get the added bonus of supporting local businesses and keeping things close to home. If you have local farm stores or small hatcheries in your area, it's definitely worth checking them out.

Especially if you like supporting local businesses.

That doesn't mean you can't buy from the hatcheries, but you need to factor in losing at least a bird per 20 purchased in transit.

That may sound harsh, but that's the honest truth of buying birds from hatcheries and having them transported through normal transport lines.

Not only do local farms provide chicks that are more likely to thrive (in most cases), but you can also see the conditions in which the chicks have been raised.

I can't stress this enough: when you're getting animals, whether it's chicks, goats, or anything else, you want to know they've been raised in a clean, safe, and stress-free environment.

No one wants to be the cause of future problems by buying from a place that doesn't treat their animals well. So, always opt for local when you can. You'll be helping your homestead, your community, and most importantly, those little chicks.

Disclaimer: There are, like every profession, unscrupulous people selling livestock. If you can get recommendations from locals, that is a great place to start. And don't be afraid to ask a lot of questions. Any honest and committed animal breeder will happily let you see their setup. They may have biosecurity requirements regarding you walking around their breeding grounds, but they will be transparent if they are genuine. Don't be scared to walk away if your gut is telling you something is off. Better to walk away than have livestock health issues later.

There is one thing to factor in when buying from a backyard hatchery versus a commercial hatchery. That is, the commercial hatcheries will vaccinate chicks.

All chicks are vaccinated at commercial hatcheries; they obviously can't afford to have disease go through their premises. However, in most cases, backyard hatcheries tend not to vaccinate as the vaccination supplies come in lots of 100 bird vaccinations at the bare minimum, making vaccinations not always viable for smaller hatcheries.

What to Look for in Healthy Chicks—Because Sick Chicks Aren't Worth Your Time

Now that you've decided on which way you'd like to purchase, it's time to learn how to pick out healthy chicks. I know, I know—they all look cute, and there is a risk you will get attached and view them like pets.

Save that emotion for your egg layers, and always remember these ones are to feed your family. You need to be smart about it. You're not just buying a cute decoration for your living room. These little fluff balls are going to grow into your future meals, so you need to make sure you're getting healthy, robust chicks.

I would always recommend visiting the hatchery or breeder before you purchase the first batch of chicks to ensure that you're getting a good quality check from that hatchery.

Here's what you're looking for:

Clear Eyes: Healthy chicks have bright, clear eyes. If you notice any crust or discharge around the eyes, walk away. That could be a sign of a respiratory issue, and trust me, you don't want to deal with that.

Clean Vent: The vent is the area around the chick's backside. You want this area to be clean and free from any signs of pasty butt. If there's anything blocking or crusting around this area, it's a sign the chick might have digestive issues.

Active and Alert: Healthy chicks are active. They should be moving around, pecking at the floor, and exploring their environment. If a chick is lethargic or seems to be hanging out in one spot; that's a red flag.

Feather Condition: Their feathers should be smooth and intact. If you notice that the chick's feathers look ragged or missing, it could be a sign of stress or poor health.

No Limb Issues: Make sure to check that the chick's legs are straight and sturdy. Any deformities or signs of weakness could cause problems later on. A healthy chick will stand and move with no issues.

Now the chicks you are looking at may not be the ones you purchase. You will likely have to place a future order, so keep that in mind. But it will give a good indication of what you are going to purchase.

Picking the right chicks is crucial for setting yourself up for success, so don't be swayed by the cuteness factor. Take a minute to observe the chicks they show you. Really consider them to ensure they meet these health standards. It'll save you a lot of trouble down the line.

The real bonus in visiting the hatchery or farmer before you purchase the chicks is that once you are satisfied that the chicks from that source are well looked after and healthy chicks, you can keep purchasing without having to go back to the hatchery or farm.

Remember that freight will not be an issue if done correctly and the boxes are kept in a climate-controlled area. Check with the hatchery how and who they freight with, and don't hesitate to ask the questions around climate-controlled freight.

Brooding and Care

Okay, so now that you've got your chicks, what's next? Time to get them settled in their new home.

A brooder is where they'll live until they're big enough to move onto pasture, and it's crucial to set it up correctly. Let's break it down so your chicks are happy, healthy, and ready to grow into those juicy, meaty birds we're aiming for.

You need to have this area set up before the chicks arrive; do not wait until the last minute to set up the brooder. It's best for the chicks that they can be taken out of their transport box as soon as possible.

How to Set Up a Brooder That Will Ensure Healthy Chicks

First things first, you need to create a brooder that's safe, warm, and comfortable for your new chicks. The good news? It's not as complicated as it might seem.

Start with a large plastic container, wooden crate, or a kiddie pool—whatever you've got on hand, as long as it's easy to clean. You'll need enough space for each chick to have about half a square foot to start with, but as they grow, you'll need to give them more room to stretch their wings. You can always add extra space by moving them to a larger brooder or a pen as they get bigger.

We use a divider board in our brooder. Then we simply move the board to give the chicks more space as they grow.

Figure 3 Meat Birds on the left and egg layers on the right

The next thing to consider is bedding. You want something that's absorbent and soft but also easy to clean. Pine shavings are my go-to choice.

They work well, are inexpensive, and they won't irritate the chicks' respiratory systems like dusty hay or straw can.

Don't use cedar shavings, though. They release oils that can be harmful to the chicks. Stick to pine or another soft, absorbent bedding.

We use a deep fill method of bedding where we don't clean the bedding out completely, instead put an extra layer of pine shavings over the top of what was already there, usually every couple of days. We do this in a continual fashion, laying an extra layer down over the top of the soiled bedding.

It does build up quite deep, but that bedding can then be used as compost after the chicks have come off it. It also saves you from having to continually clean it out and re-lay bedding, which is disruptive and stressful for the chicks.

It's much easier just to put a little over the top to soak up any moisture that's in the pine shavings and keep the smell down. This gives the chicks a clean and dry environment to live in.

Now, let's talk about heat. Baby chicks can't regulate their body temperature, so they need a heat source to keep them warm and cosy. This is where a brooder plate or heat lamp comes in handy.

You'll want the temperature to be about 35°C (95°F) for the first week, then drop it by 3°C (5°F) each week until they're ready to move out of the brooder and onto the pasture.

Figure 4 Day olds under a heat lamp

Place the brooder plate or heat lamp over one section of the brooder so the chicks can move in and out of the heat as needed. This way, they can regulate their temperature by moving closer to or further away from the light.

If they're huddled together directly under the light, it means they're too cold. If they're avoiding the heat and sitting at the far end of the brooder, they're too hot. Keep an eye on them and adjust the temperature as needed.

The brooder plate that we use (we're actually looking at stocking on our Amazon store very soon) has a set temperature that doesn't need to be adjusted.

It has a low and high setting and extra legs to raise it up higher to allow the chicks to have a slightly cooler environment by giving more headroom. We found that this is much better than a heat lamp for power usage, as we are off-grid, and poses zero fire risk.

Figure 5 Home-made chick feeder

What to Feed Your Chicks

Chicks should be fed commercial chick starter crumble.

This is made up of crushed pelleted food that contains the correct ratio of proteins, carbohydrates and fats that chickens need to grow. It also contains vitamins and minerals that are essential for healthy development.

Some people may wish to make their own feed up, especially to be truly self-sufficient, however, I would caution against this.

Buying commercially made feed takes all the guess work out of something that could end in disaster from a health perspective.

The correct percentage of protein, which is the most important factor for growth is as follows

- starter – minimum 20% protein
- grower – 15–16% protein

If you unsure of what is available in your area, please speak to your local produce store or tractor supply store. Staff there are usually very knowledgeable on the subject of feeds.

I'd would also like to touch on the subject of medicated chick starter. Some people are averse to feeding their chicks anything that has medicine added to it. To be honest; I would prefer to feed them the medicated chick starter than have any kind of loss in numbers of chicks.

If using hybrid meat birds, the medicated starter only gets fed to them while in the brooder, which is less than half of their life, usually only 3 weeks out of 8 weeks.

I have looked at studies to see how long the medication stays in their system, and I believe that it is still worth feeding it to them for the first three weeks of their life to ensure health. The half-life will be out of their system before you dispatch and process the bird.

That being said, I also am completely okay with people who don't medicate or vaccinate their animals that are going to be used for human consumption.

While not wanting to go down the conspiracy theory paths, I also believe that we probably put way too many chemicals into our bodies and understand somebody's choice not to add any more chemicals to the food that they're eating.

Remember to go and speak to your local produce store or tractor supply about what type of feeds are available in your region. They will let you know what medicated and unmedicated chick starters are available. Unmedicated is usually referred to as a natural chick feed.

If you do decide to go down the route of not having a medicated chick starter, just be aware that there are certain things that can develop in chicks that can cause you to lose your entire batch of chicks.

Coccidiosis is the main reason for feeding a medicated chick starter. This parasite can wipe out your entire batch in no time; however, it is a risk that some people are prepared to take. Chicks do develop an immunity to Coccidiosis after exposure, and only if they are in good health.

Daily Care Routine: Heat, Bedding and Feed

Alright, so you've set up the brooder, and the chicks are happily chirping away. Now, it's time for the daily care routine. Sure, it sounds boring, but trust me, it's crucial. This is how you'll ensure your chicks grow into healthy birds, and that means good food, fresh water, and keeping everything clean.

Figure 6 A brooder plate

Start by checking the temperature of the brooder at least once a day (preferably twice a day) to make sure it's at the right level. The last thing you want is for your chicks to get too hot or too cold. A simple thermometer can help you monitor the temperature and keep things in check.

The ambient temperature doesn't have to be sitting at 35°C (95°F), but if your thermometer is close to the heat source, then it should be pretty close to that.

As I said earlier, you can monitor your chicks by seeing if they are hovering around the heat source or if they are far away from your heat source; that will let you know if the heat source is either too high or too low.

Next, feed your chicks a high-quality chick starter feed. This is designed to give them the protein, vitamins, and minerals they need to grow strong. You'll want to keep their food and water containers clean and make sure they always have access to fresh water.

Chicks drink a lot, so make sure to keep the water clean and topped up. This will help them grow strong and healthy without the risk of dehydration.

Keeping Chicks Cozy and Safe

As your chicks grow, you'll start to see their personalities shine through. Some will be little adventurers; others will be shy and huddle together for warmth. I've had a batch of Cobs who had a couple of aggressive ones! Every time we fed them, we'd get a little peck on the hand, and they would fly at us. And they were only a couple of days old, feisty little buggers.

Remember, it's your job to make sure they stay cosy and safe; you need them to grow to their full potential. Check the bedding every day to make sure it's dry and clean. Chickens can be messy, and wet bedding can lead to health issues, so don't let that slide.

If using the deep fill method, top it up anytime you notice a smell or it appears soiled to keep everything fresh. If you can smell it, then it needs to be added to.

When it comes to food, don't be tempted to offer them treats too soon. Chick starter feed is all they need in the beginning.

As they get older and start transitioning to pasture, you can start offering them a little variety, but for now, stick to the basics.

Regarding scratch mix, which I call "Chicken Candy" do not feed it to growers, or even layers for that matter, it's not the best feed and should only ever be a small treat.

Chapter 3

Transitioning to Pasture

Alright, folks, the time has come. Your chicks have grown up enough to move out of the brooder, spread their little wings, and start living their best lives on pasture. This is where the magic really begins. And trust me, this is where you'll start feeling like a true homesteader—watching your birds go from cute little fluff balls to robust, pasture-raised meat machines.

But before you throw open the brooder door and let your chickens roam free like it's some sort of chicken utopia, there are a few things you need to keep in mind.

Transitioning your chickens to pasture isn't as simple as setting them loose in the yard and hoping for the best. Nope, there's a bit more planning and prep involved to ensure they stay safe, happy, and healthy while also giving you the delicious, nutrient-rich meat you're aiming for.

In this chapter, we're going to walk through all the key steps for transitioning your chicks to pasture. We'll talk about setting up the right space, how much room they really need (spoiler alert: it's not as much as you think), and how to move them to their new home without turning it into an episode of "Survivor: Chicken Edition." If you live in an area with predators, it could become!

Preparing for Pasture

Okay, first things first: Let's talk about what needs to be done before you actually move your chickens to pasture. You don't just want to throw some chickens out in the open and hope for the best. This is where the setup really matters, because if you get it right, your chickens will thrive, and you'll be living the self-sufficient dream.

Fencing, Shelter, and Pens: You Don't Want Your Chickens Going on an Adventure

Chickens love to roam, and if you're anything like me, the last thing you want is for your birds to be roaming into your neighbour's garden or, worse, becoming a tasty treat for a fox, coyote, eagle, or hawk.

So, strong fencing is essential. And I'm not just talking about the stuff that'll keep your dog in the yard; we're talking about solid, chicken-proof fencing.

While I keep my egg layers in a fenced area, for my meat birds, I prefer the movable chicken tractor style. The Handy Helper built mine based on the Joel Salatin from Polyface Farms style of meat birdhouse.

Figure 7 Our homemade chicken tractors

It's basically a portable chicken house with no floor and you can move it around your pasture, so the chickens get fresh ground, grass and bugs every day.

The beauty of this setup is that it not only gives your chickens access to fresh pasture, but it also helps with fertilising the land in a controlled way. You're basically giving your land a little bit of love while your chickens do the heavy lifting. Chicken manure is so good for the pasture, and fresh grass is so good for your meat birds.

The real issue with raising any livestock for human consumption is managing parasite loads. Paddock rotation or ensuring that the livestock always have access to fresh pasture means that you don't have to concern yourself as much about their parasite load.

This means that the animals that you're using for human consumption will not have to be as medicated or chemically "cleaned out" as an animal that would be, say, in a feedlot and not having access to fresh pasture.

It is the main reason that paddock rotation works so well for livestock, and animals are just healthier.

If you choose to have a conventional fenced-off area, then the fencing you choose needs to be sturdy and secure. It should be tall enough to prevent your chickens from flying out (most meat chickens can't fly very far, but they can manage a short hop), and it should be buried into the ground or electrified to prevent predators from digging under.

The mesh should be small enough to keep out even the most determined raccoon or fox, so thin welded wire or hardware cloth, not soft chicken fencing (that stuff's more of a suggestion than a barrier).

The only time soft chicken fencing will work is if it is permanently electrified. You should check that it is still working daily, as it can short out in long grass and become ineffective.

As for shelter, if you are using a chicken tractor, then half of the "box" should be enclosed to give shade as well as protection from wind and rain. If using a fenced area, then you need to have a house within the area that does all the same things.

It doesn't need to be fancy, and at Mojo Homestead we are big advocates of using second-hand materials rather than shiny new ones. You don't need to overdo it, but a simple roof is enough.

We have found our hybrid meat birds do not roost like normal chickens, and as long as you move them daily, they will never be sleeping in their own manure.

How Much Space Do Chickens Really Need? (Hint: Not As Much As You Think, But Enough to Keep Them Happy.)

Now, this is where things can get a little confusing. You might be thinking that your chickens need acres of pasture to really thrive.

And while they do need some space to roam and forage, you'd be surprised by how little space they actually need to stay happy and healthy.

While overcrowding them can lead to stress, pecking issues, and a bunch of other problems, they don't need a lot.

So, let's talk about how much space is just right.

I hate it when people list "chickens just need .18 m2 per bird" and expect me to translate that! I just can't visualise that area.

Joel Salatin's tractors are either 2.4 m by 3.6 m (8 by 12 feet) or his larger version was 3.0 m by 3.6 m (10 by 12 feet). Joel says that his larger ones aim to house 75 birds.

Our homemade version is 2 m by 2 m (roughly 6.5 feet by 6.5 feet), and we have housed 20 birds in it, moving it daily without a problem. Honestly, when considering a fenced area, consider the same! Meat birds don't need a lot of room as they will "forage" a lot less than egg birds.

Egg laying birds will generally need more, but we are not going to get into that in this book. However, don't stress too much about getting this exactly perfect. Chickens are social creatures, and as long as they have room to roam, peck, scratch, and run, they'll be perfectly content.

The key to happy chickens on pasture is giving them access to fresh grass, insects, and bugs. The more they can scratch around for their own food, the healthier they'll be, and the better your meat will taste.

A little crowding isn't the end of the world, but keep an eye on them and adjust if they start acting stressed or agitated, then consider smaller batches or a second tractor.

Moving Chicks to Pasture

So, you've got your pasture all prepped, your chicken tractor is looking like a little slice of chicken paradise, and now it's time for the big move.

But hold up—before you open that door and let your birds fly into their new home, there are a few things to consider. Timing, technique, and some serious patience are required to make this move as smooth as possible.

Timing and Techniques: You're Not Just Opening a Door and Letting Them Run Free

Figure 8 Chick with not enough adult feathers

The last thing you want is a bunch of confused chicks darting in every direction when you move them to pasture. It might be tempting to open the door and watch them run wild, but let's be real: that's not going to end well for anyone.

Timing is key here. Before you move your chicks to pasture, make sure they're old enough to handle the transition. If they still have fluff, they are not ready.

Figure 9 Chicks with enough adult feathers

If they're still too small or haven't developed enough feathers, they won't be able to regulate their body temperature as well, and they might get cold or stressed. I usually wait until they're around 3 weeks old.

Hybrid meat birds will have their adult feathers by then, heritage breeds, you may have to wait until closer to 6 weeks.
Once they have mostly normal feathers and little to no fluff, they are ready to go.

In our brooder house, we have a small outdoor fully enclose section that we let them forage in during the day only from 2 weeks for hybrids, and 3-4 weeks for heritage. This is to help acclimatise them.

We also move them in bundles in a secure container. These birds will generally not be tame, and while hybrids are not great flyers, heritage breeds are, so always move them securely. Don't stress if you don't have a "proper" crate. We have used cardboard boxes or dog crates to move our chickens. It just needs to be secure during the move.

Figure 10 Meat hybrids at 3 weeks moving to chicken tractor

Make sure water and food is in the tractor before putting the birds in, this helps them adjust to their surroundings immediately.

We also like to move them in the morning because we then move the tractor each morning, so they get 24 hrs on each piece of pasture.

While hybrids are usually sedate and do not have "pecking order" issues.

We have notice that heritage birds do! Likely because most of them are roosters, we will keep any pullets as they develop, and we separate them if they are obviously not a rooster.

Don't stress if the pecking order fights happen, most flocks will sort themselves out pretty quickly. Once you have all the birds into the tractor, it's important to watch for 2 things.

Overcrowding - You notice they may lay on top of each other. If this happens, it could be too many birds in the cage or not enough protection from the elements forcing them into one corner. You'll need to deal with that.

Turning pasture to dirt too quickly - If you see this start happening, then you'll need to move the tractors every 12 hrs instead of 24hrs. It means you've got some good appetites on board, and likely quick growers, and that's a good thing.

It might take them a little while to adjust to their new environment. Some might be hesitant at first, but as soon as they realize that fresh grass means fresh snacks, they'll be all over it.

What to Feed Pasture Raised Chickens

As I mentioned in the brooder chapter, commercially made feed is a bit of a "no-brainer" for me. It contains everything that growing chickens need for health.

Grower food is a pelleted mix of everything they need, like starter crumble however in the larger pelleted version.

Protein percentages should be between 15-16%. There is no need to feed scratch mix, it's expensive and they will fatten up fine without it. If you wish to supplement their food with scraps or excess produce, you can but it is not essential.

Benefits of Pasture-Raised Chickens: Healthier Birds, Better Flavour, and Yes, It's Self-Sufficient!

Moving your chickens to pasture isn't just about giving them more space to run around (though that's definitely a perk). There are real, tangible benefits to pasture-raising your birds, and trust me, they're worth every bit of effort you put into setting it up.

First off, pasture-raised chickens are healthier birds. By foraging for their own food, they get a more natural diet full of grass, bugs, and worms—things that boost their immune systems and help them grow strong and hearty.

Chickens that are raised on pasture are less likely to suffer from health issues like leg problems, and they're far less stressed than those kept in small, cramped conditions.

But the biggest benefit? The flavour.

Pasture-raised chickens are known for their superior taste. Like grass fed beef, the diet of grass combined with omnivore insect snacks contributes to a richer, more flavourful meat.

And because they're raised naturally, you don't have to worry about weird additives or antibiotics sneaking into your meals. You're getting clean, wholesome, and incredibly flavourful chicken, just like nature intended.

Plus, it's a step towards self-sufficiency. When your chickens are out on pasture, you're not only giving them a healthier, more natural lifestyle, but you're also reducing your reliance on store-bought feed and other supplies.

You're closing the loop on your own food production, which is the heart of self-sufficient living.

Now I do want to add that from personal experience we have noticed that while all pasture raised meat birds taste better than commercially raised ones. My family noted that the heritage breed (Rhode Island) that we raised have a much stronger taste than the hybrids (Cobs).

Wrapping It Up

Getting your chickens transitioned to pasture is a big step, and while it takes some time and effort, it's also one of the most rewarding parts of the process. Watching your birds move from the brooder to fresh pasture is a satisfying moment and knowing that you're providing them with the best possible environment to thrive makes it all worthwhile.

By this point, you're not just raising chickens for meat—you're living the homesteading dream. You're part of a movement that's reconnecting with the land, taking control of your food sources, and ensuring that what you put on your table is healthy, delicious, and full of love (and not to mention, a lot of hard work).

Keep it up, and soon enough, you'll be enjoying the fruits (or rather, the meat) of your labour. Just remember as you move forward, always keep your chickens happy, healthy, and safe on that lush pasture.

The payoff in flavour and self-sufficiency will be worth every bit of effort you put in. Now go ahead, open that door, and let those chickens live their best life.

Chapter 4

Finishing and Dispatching

Alright, so you've got your chickens living the good life on pasture, and now you're getting close to the big moment. I won't lie to you—this chapter is a little more serious than the previous ones, but it's also one of the most important. Whether it's your first time or your fiftieth, dispatching your chickens is a moment that deserves your full attention. This process isn't about being squeamish or rushing through it; it's about doing what needs to be done with respect, care, and intention.

But before you start getting too nervous, remember this: You've raised these chickens with love, and now you're going to take responsibility for providing your family with a nourishing meal. This isn't just about eating; it's about becoming self-sufficient, taking control of your food, and doing it in a way that's humane and respectful to the animals you've raised.

Also, I believe it is important to note that if you cannot dispatch your own chickens, do not beat yourself up about this. It doesn't mean you have failed or are not a true homesteader. Never let anyone tell you that.

Dispatching animals used to be commonplace in the past, but it is not that way now. If you need assistance to dispatch, that is fine. I cannot stress how important it is to be gentle on yourself regarding this.

Look to your family or community in the first instance. I honestly believe that most people will adapt and learn dispatch once they have seen it firsthand. It is much easier to do that your imagination will convince you.

That all being said, let's dive in and make sure that your chickens get the proper treatment they deserve—both at the end of their lives and in the final preparations.

Finishing Chickens on Pasture

Before we get to the dispatch part of this whole operation, we've got to talk about the "finishing" phase. This is the part where you make your chickens the best they can be in terms of flavour and quality. And it's not just about throwing feed at them and hoping for the best.

No, this part of the process is critical because you're feeding your chickens not just for weight but for flavour. That's right, we're talking about the kind of wonderful chicken flavour that you can almost taste just thinking about it.

Just like grass-finished beef, the last growing phase is important.

What to Feed Them When They're Finishing Up: You're Feeding Them for Flavour, Not Just Weight

When it comes to finishing your chickens, it's a matter of timing.
You want to have them grow as big as they can but don't try to stuff food into them to get to that point. Just Continue to feed the grower pellets.

There are versions of finisher pellets however I have never used them. They are more expensive where I live and seem to contain all the same ingredients.

For the last few weeks before processing, your chickens should be getting the 15-16% -protein feed.

While your birds have been foraging on grass, bugs, and other natural food sources that cannot supply them all the nutrients needed, so ensure they are getting enough pellets.

You have a choice with grower pellets as to medicated or non-medicated. It is a very personal choice; however, if you are concerned about additives, you should choose unmedicated. We have used the unmedicated for all our meat birds with no issues regarding coccidiosis.

Now, you might be tempted to think that the more you feed them, the bigger they'll get. But that's where you've got to watch yourself. You don't want to overfeed and end up with a bird that's too fatty or one that's growing too quickly.

Overfeeding leads to poor meat quality—think of it like the difference between a free-range chicken and one that's been pumped full of growth hormones.

Too much weight too quickly can mess with their texture and flavour, so feed them enough to finish them off nicely without going overboard.

For meat-specific birds, overfeeding can also cause problems; however, that is usually in the first few weeks. Towards the end of their growth, it is hard to overfeed, but like with all animals, you will have some that gorge. Don't be tempted to go crazy with the feed.

You should ad lib feed, but as a rough guide, a grower will eat about 6 kg (13 pounds) of pellets each between week 4 and weeks 8-9.

A good approach is to offer them fresh pasture during the day, but supplement that with a balanced chicken feed at night to make sure they're getting enough of the good stuff to put on some weight and flavour. You're aiming for a balance between nature and nutrition.

Monitoring Their Growth: How to Know When It's Time to Say Goodbye to Your Meat Birds

This part can be tricky, especially if you've never raised chickens for meat before. How do you know when it's time to process them? You don't want to let them get too old; some people believe the meat will get tough. But you also don't want to jump the gun and harvest them too early.

My bigger concern is spending money paying for feed when it is not growing a bigger bird.

Most meat-specific birds will slow dramatically in their weight and size growth at a certain point, with no amount of feed changing that. It would be a waste of feed to continue to in order to get a 'bigger bird'.

The easiest way to gauge when your birds are ready for processing is by monitoring their weight and growth.

Most meat chickens, like the Cornish Cross, will be ready at around 8 to 9 weeks, while slower-growing breeds like the Freedom Rangers may take up to 12 weeks. You want your birds to be at least 2 kg (4 to 5 pounds) at the time of processing.

We have had batches that were closer to 3.5 kg on average (7.5 pounds) but don't get too hung up on the numbers.

It's more about how they look and feel. If they're plump, have full breasts, and are moving around energetically (but not too energetically—remember, they've been growing fast), it's time.

Another thing to consider is their behaviour. When chickens are ready for processing, they'll start slowing down a bit, especially as they near the end of the growing period.

You'll see them eat a little less, and they'll start to spend more time resting.

This is normal—it means they've reached maturity. So, when you see your chickens slowing down and they've hit their target weight, it's time to start prepping for dispatch.

Be mindful; some meat birds will get too big for their bodies to handle. We have only had it happen once; we lost a bird the night before dispatch. It appeared to have simply had heart failure.

Preparing for Dispatch

So, it's time. The chickens are big and healthy, and they're ready to be processed. Now it's time to set up your dispatching area, because this is the step where preparation is absolutely key.

When you're working with live animals, things need to go smoothly, and you want everything to be as calm and organised as possible. Trust me, the smoother this process is, the better it'll be for you, your birds, and the quality of your meat.

Taking the Birds Off Feed

At least 24 hours before you plan to dispatch, take the birds of grass, and give no food. Provide fresh, clean drinking water for them but no grain or pellets.

You want the birds to flush their systems of any food. Trust me when I say this will make the process much cleaner.

We have trialed 12-hour fasting, and just about all the birds still had some vegetable matter in their stomachs. If you can't do this, it's not the end of the world, but it does make disassembly a little bit messy.

The Tools of the Trade: Kill Cones, Scalpel, and Everything You Need to Do This with Respect and Care

Now, you're probably thinking, "Okay, what exactly do I need to dispatch these birds properly?" First, you need the right tools.

Here's a quick rundown of the essentials:

Kill Cone: This is your best friend when it comes to dispatching chickens. A kill cone holds the bird in place while you do the job. It keeps them calm, it prevents flapping, and it ensures that the bird is positioned correctly for a swift, humane process. You can buy one, but if you're a

DIY enthusiast like me, you can easily make one using some mesh or sturdy plastic.

Figure 11 Metal kill cone fixed to the wall

Figure 12 Large scalding pot on gas burner

Large Pot on Heat: You will need a scalding pot that is set up on heat ready to go. We use an old disused preserving pot that is about 20 litres. It is set up on our old BBQ gas heater, and we have a thermometer sitting in the water.

You need to be able to keep the water at a constant temperature of 60°C (140°F) throughout the process, so set this up before you start. A good tip is to add a little dishwashing detergent to the water as it helps loosen the feathers.

Sharp Knife or Scalpel: You're going to need a sharp, clean knife to quickly and efficiently sever the bird's carotid artery. A dull knife is a no-go—it's messy, it stresses out the bird, and it makes the process take longer.

A scalpel works well because of its precision, but a sharp kitchen knife will do just fine.

You will also need a knife to disassemble. I use different knives because my dispatch knife has to remain super sharp, and my disassemble is used for cutting through bone. I have even used gardening shears to remove feet; it helps keep your knives sharp.

Figure 13 Protective apron, gloves and gumboots

Apron and Protective Clothing: I have a heavy rubber apron that I bought from Amazon; it's very long and I've never damaged it, so it must be tough.

I also wear gumboots, as you will have a lot of water around, and wet feet are just uncomfortable.

I also have super thick heat-proof gloves for the scalding process. I usually put a cap on for the process; if you have long hair, there is nothing more frustrating than hair in your face when your hands are covered in chicken blood and guts.

Buckets and Tubs: You'll need a couple of tubs or buckets. I have one under the kill cone to collect blood. I find the flies get a bit hectic if I don't keep it off the ground.

I also have one for the heart, liver, and necks, as they are used for pet food. This one has a lid to prevent flies from getting into it.

The last bucket is for waste offal—all the things that are just going to compost. You can choose if you lid this bucket, but as I'm in Australia, the flies make me lid this, but I don't care if it's uncovered while I'm disassembling.

They need to be buckets that can be sterilised as well. You don't want to be scrambling around for containers in the middle of the process, so have them set up and ready to go before you start.

Plucking Setup: Once your bird is dispatched, you'll need a plucking station.

If you're using a homemade plucker (cheers to you for being handy!), make sure it's set up and ready to roll.

Otherwise, be prepared to pluck by hand, which can take a little longer but isn't impossible.

Figure 14 Homemade plucking machine

This is our plucking machine that the Handy Helper made. It uses an old cement mixer motor, a recycled washing machine tub, and rubber fingers bought from Amazon. He then bolted in to a frame to hold it steady and off the ground. We use an old garden pot to collect feathers in.

Esky or Cooler: You need somewhere to put the carcasses once disassembled. It's impractical to stop and wrap every bird as you do them.

So set up an esky or cooler bin with cold water and ice; you want the carcass to cool as quickly as possible. Once the bird is gutted and cleaned, just place it in and shut the lid; this will stop any contamination.

Setting Up a Safe, Clean Dispatch Area: You Want Everything to Go Smoothly (and Cleanly)

Now that you've got the tools ready, let's talk about setting up the space where the magic happens. First off, you'll want a calm, clean area—somewhere that's far from distractions, kids, or overly curious pets.

Figure 15 Second-hand glass top table with drain hole

I recommend setting up near a place where you can clean up easily, and if possible, in an area where you can easily dispose of waste. A large table or countertop will work well for gutting and processing, and you'll need a separate area for plucking. We use an old patio table that has a glass top. That means I can sterilise it easily.

You will need access to a hose that can be easily turned on/off; a spray trigger works best for me.

Set up your kill cones in a safe, secure spot where you have plenty of space to move around. You don't want to feel rushed during this process, so give yourself room to work.

Also, make sure that you've got everything ready to go before you start—tools, buckets, water, gloves, etc. Having a clean, organised setup will make the entire process go much smoother.

I also find it a good idea to switch your phone to "do not disturb" mode and let anyone know who may need to contact you in an emergency that you will not be available.

You really don't want distractions the first time you are dispatching; it's important to remain focused.

The Dispatch Process

Alright, here we go. It's time to dispatch your chicken. The next steps are crucial—this is where the respect, care, and attention you've shown your chickens up until this point really comes into play. I'm not a religious person, but some people feel it's important to say either a blessing or a prayer before they commence. If this is something you feel is important, now is the time to do that.

Step-by-Step: Using a Kill Cone to make for a Calm, Efficient Process

1. Place the Bird in the Kill Cone: Gently and confidently place your bird headfirst into the kill cone. The bird should be calm but alert, with its neck exposed and ready for the next step.

2. Make the Cut: With your sharp knife or scalpel, make a clean cut along the carotid artery and jugular vein. You want to be swift and precise. A clean cut will ensure a quick and humane death, allowing the bird to bleed out efficiently.

3. Let the Bird Bleed Out: This is a critical part of the process. Allow the bird to bleed out for a few minutes—this will ensure that the meat stays clean and of the highest quality.

Don't rush this part.
Take a deep breath and let the bird pass peacefully.

Figure 16 Placing the chicken in the kill cone

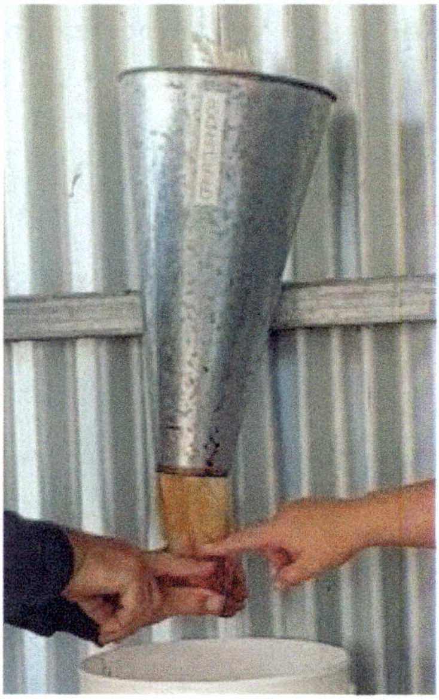

Figure 17 Ensuring the neck is exposed correctly

Figure 18 Holding the head down to allow bleed out

Ethical Considerations: We're Homesteading, Not Running a Factory Farm

As you work through this process, always keep in mind the respect and responsibility you have as a homesteader. This is your food, your livelihood, and a living creature you've raised with care.

By dispatching your chickens humanely and efficiently, you are honouring the work you've done and providing your family with a nourishing meal. Be mindful and aware throughout the process—this isn't about rushing or being squeamish; it's about handling the task with the respect and care it deserves.

Post-Dispatch Cleanup: Because No One Likes Dealing with Messes

Some people like to give the area a quick spray of water once the bird has been removed from the kill cone.

Don't get hung up on it because you still have a bit of work to do yet. That being said, blood can be hard to clean once it is dried, so I usually just give a quick spray with the hose to dilute any blood.

Take a moment to give yourself a pat on the back if that was your first dispatch.

It's honestly the part most people struggle with, but if you are set up correctly, it is the quickest section of the entire process.

Now we'll move on to the next phase: processing, plucking, and breaking down the meat.

Wrapping It Up

Dispatching your chickens is one of the most important parts of the process, and it deserves your full attention. With the right tools, a calm environment, and respect for the life you're taking, you'll be able to process your chickens efficiently and humanely. You've raised these birds from day-old chicks, and now you're providing your family with the fruits of your labour.

Take a moment to appreciate what you've accomplished—it's no small feat.

Now, on to the next chapter, where we'll talk about processing, plucking, and preparing your birds for the table. You're getting closer to that delicious home-raised chicken, and trust me, it's going to be worth every bit of effort you've put in.

Keep it up, because this is where the hard work turns into something incredibly satisfying.

Chapter 5

Processing the Chicken

Congratulations! You've come a long way since starting with those tiny day-old chicks arriving at your homestead. You've raised them healthy, transitioned them to pasture, and managed to dispatch.

The next step is processing, and I'm not going to sugarcoat it—this is the part where things get a little messy. But it's also the part where you truly see the benefit of your labour, and trust me, it's worth every bit of effort.

In this chapter, we'll break down the entire processing process, from feather removal to storing your meat. Remember, the key here is to approach it methodically. You've done a lot of hard work up until now, so let's get in and make sure the final steps are as smooth and efficient as possible.

And let me reassure you—this isn't as daunting as it sounds. Once you get into the rhythm, you'll start to feel like a pro. So, let's get started.

Feather Removal

The Best Techniques for Getting Feathers Off Your Birds

Feather removal might sound like the kind of chore you'd want to avoid, but I promise, it's not as bad as it seems. Once you get the hang of it, you will become super efficient. The trick is to use the right technique, and let me tell you, scalding is your best friend. Correct scalding makes the whole process faster and you end up with a perfectly plucked bird.

What does it involve? Scalding involves dipping the chicken in hot water to loosen the feathers. The pores will open, allowing the feathers to easily slide out of the skin. It can be a bit of a learning curve, so don't get frustrated; just follow the instructions and make a mental note of what works best.

Here's how to get it right:

1. Get the Water Temperature Just Right: The key to scalding is the temperature. Too hot, and you'll cook the skin (which is a no-no); too cold, and the feathers won't come out easily. Aim for about 60°C (140°F). To test it, dip your hand in the water. It should feel hot but not burn you. We use our heavy-duty gloves as our children help with this process, and while I have milkers' hands (I can tolerate pretty hot water), they don't. So, if you are hot water sensitive or kids are helping, be careful not to burn yourself.

2. Dunk the Bird: Once you've got your water at the right temperature, dunk your chicken in for 30 to 60 seconds. You'll notice the feathers starting to loosen up. Don't leave the bird in too long, though, as you don't want the skin to cook. Now this is the tricky part. You might need to dunk 2-3 times! Test if ready by pulling on a feather; if it comes away easily, then you are good to go. If it requires a lot of effort to remove the feathers or they don't come out at all, then you need to keep dunking.

3. Pluck Away: After scalding, take your bird to your plucking station. If you're plucking by hand, see below for more detail. If you've got a plucking machine (a homemade one is a great investment), it'll do most of the heavy lifting for you, leaving you with less work and more time to focus on other tasks.

Preparing for Plucking: A Little Prep Work Goes a Long Way

Before you dive into plucking, make sure you have everything ready.

If you are only doing a couple of birds, hand plucking can work. If you are processing a few, I recommend machine plucking. Either hire one or build/buy one; it is worth the investment. Having hand plucked as a kid, I will never go back to that. Even with a machine-plucked bird, you still have to remove a few stray feathers, but that is nothing in the greater scheme of things.

The way the machine plucker works is by spinning the bird while the rubber fingers pull at the feathers the same way you would pull the feathers out by hand. The difference is the machine is pushing the bird past lots of rubber fingers, making the feathers pull out really quickly. The rubber fingers can be purchased on Amazon if you are a DIYer.

Machine Plucking

1. We turn our machine on with the scalded bird already in there, but you can hold the bird and gently drop it into a machine once on. Just be careful that it doesn't grab your hand and pull you in.

2. As soon as it's on, take your hose on a spray setting and blast the bird with cold water; this helps the rubber grab the feathers and helps prevent skin tearing.

3. Never walk away while this process is happening. You want to stop the machine as soon as the majority of the feathers are out. If you keep it going, it will tear the skin.

4. Repeat the process if, after stopping, there is still a clump of feathers in the bird.

Notes:

You can remove the head and feet prior to plucking. I prefer to leave the legs on, as it's easier to grab the carcass by the legs during this process.

The skin can tear, and while this is not ideal, it is not the end of the world. Unless you are trying to sell whole chickens, appearance isn't the most important thing, so keep that in mind.

Remove the feathers from the machine as they build up; they make great compost; just bury them deep to allow breakdown.

Hand Plucking

I don't recommend this unless you only processing a couple of birds. It is time-consuming.

1. After scalding, hang the bird's feet from something solid at a height that is right for you. Too high and your arms will ache; too low and your back

will. A rotary clothesline is a great height for most. Lay down a tarp or plastic to catch stray feathers. I find having a large bucket with a trash bag liner underneath the bird is best. If you poke a few holes in the trash bag, it won't collect with water. Alternatively, if you are avoiding plastics, use a bucket to collect everything. It makes great compost; just bury it deep.

2. Bare hands work best; I have tried gloves, but they get slippery. Using your fingers, grab chunks of feather close to the skin and pull in the same direction as the feather comes out. If the bird is properly scalded, you should not have to pull hard. It will be messy, and feathers will stick to your hands; don't get too paranoid or you'll spend a lot of time washing your hands.

3. Work in sections, starting with the wings and tail feathers—they're the easiest to remove. From there, move on to the larger body feathers. Take it slow, and don't rush it. Like I said before, it is time-consuming but not impossible. Once you have got all the feathers out, give the bird a rinse with your spray hose and wash your hands at the same time.

Processing the Carcass

Gutting and Breakdown: From Bird to Meat

Now that your chicken is plucked and ready to go, it's time to break it down into the usable meat that will feed your family. This is where you will see your hard work transformed into your food supply and your path to breaking up with the supermarket.

Here's how to go about it:

1. First, I remove the head and feet (if not already removed in earlier steps); these go in your waste bucket, unless you plan on using the feet, in which case place in the keep bucket.

2. Next, remove the gland that sits just above the tail area; use the main tail feather point to find it. It's also known as the preening gland, and it is removed because it can have an unpleasant taste. You will see it easily, as it is more yellow than the below photo shows.

Figure 19 The gland is under the pin feather

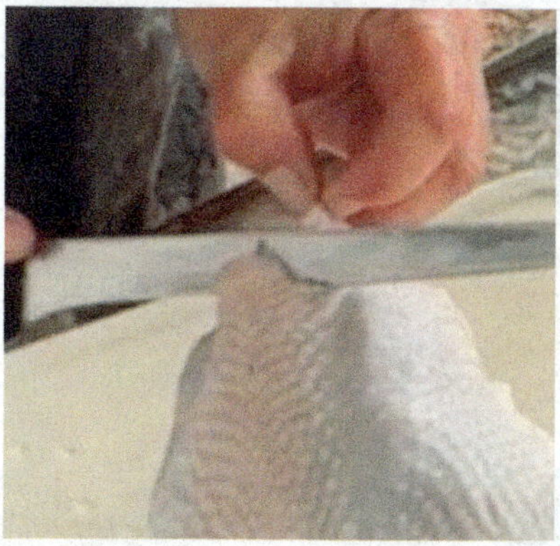

Figure 20 Cut on a 45 degree at front and back

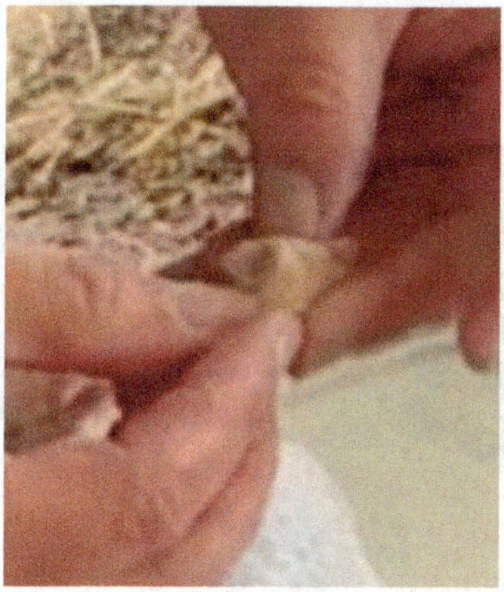

Figure 21 The gland tissue will be yellow

3. Next, with the bird on its back, you want to separate the windpipe and croup from the carcass. It is held in place by slippery connective tissue, so it can be difficult to grab. You will need to get a good hold on the croup to pull it free. Don't try to pull it out of the bird; you just want it loose. This way, when you remove the innards, it will come free via the vent end of the carcass.

Figure 22 Separate the windpipe/croup from carcass

Figure 23 This separation will require some strength

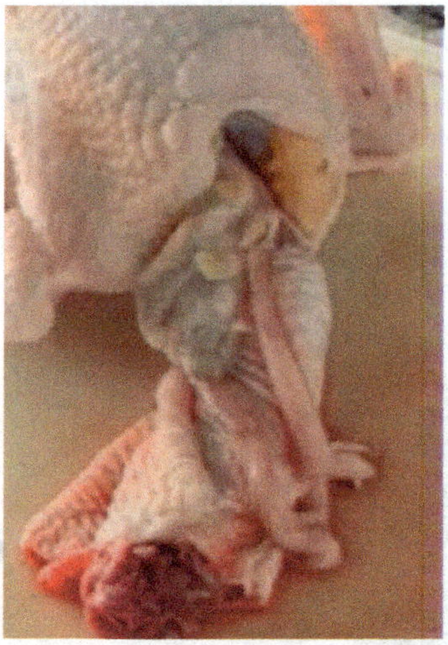

Figure 24 Don't pull it through the next, it will come out the other end

4. Now flip it over onto its back with the vent easy to access and start by making a small incision around the vent (the area around the bird's backside).

Be careful not to puncture the intestines. You want to slice just deep enough but not all the way through. Make an incision, then using your hands, tear the opening big enough to reach your hand inside. Using your fingers, push the organs free of the carcass walls. Loosen them both sides, front and back, before attempting to gently pull out the intestines and other internal organs out through the opening.

You'll need to remove everything, including the lungs, heart, and gizzard (croup). The windpipe and croup that you freed earlier should come all the way through with the rest of the innards.

Don't rush it—take your time to ensure everything comes out cleanly.

Figure 25 Make the incision above the vent, being careful not to go too deep

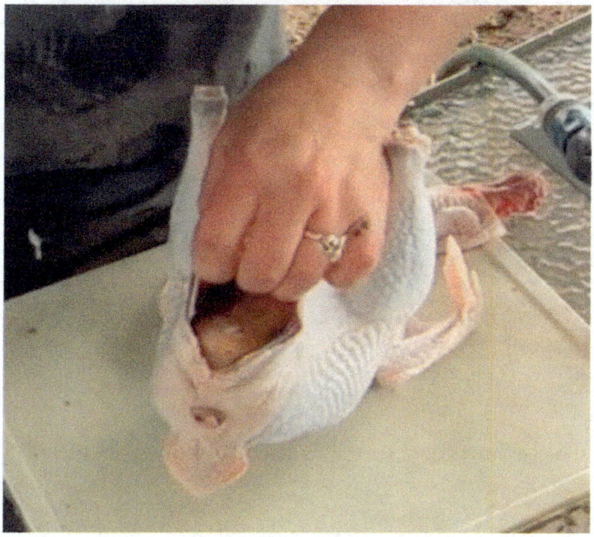

Figure 26 Instead of cutting, tear the opening wide enough to fit one hand into the cavity

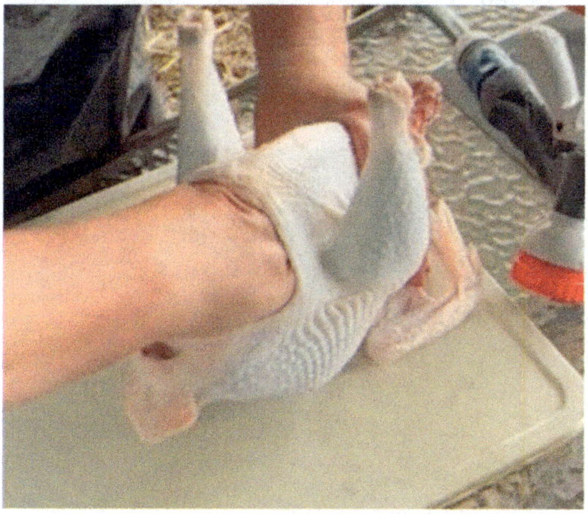

Figure 27 Using one hand separate the organs from the cavity wall

Figure 28 Try to remove the organs without puncturing any

The lungs can be hard to remove as they sit closest to the rib cage.

You can buy a special tool to scrape them out, but I have always been able to get them with my hands. If you are really struggling, a metal fork would likely work just as well.

Below are the different organs.

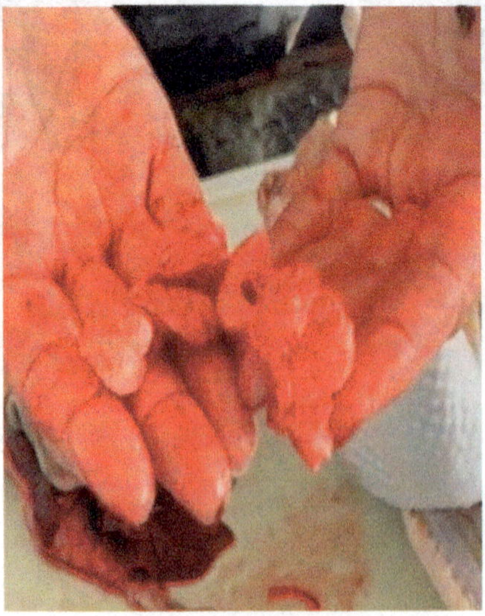

Figure 29 The lungs

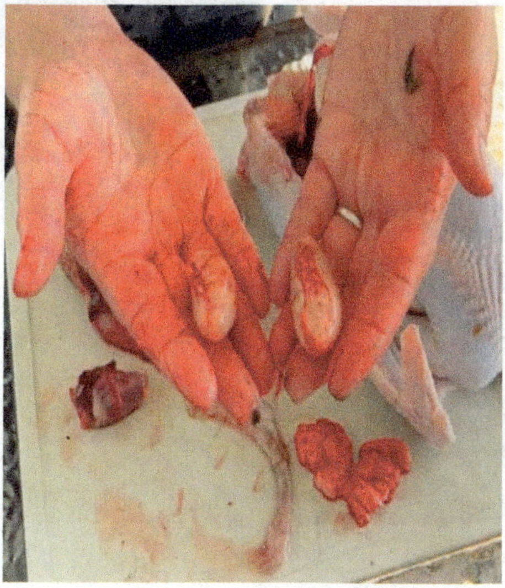

Figure 30 The testes

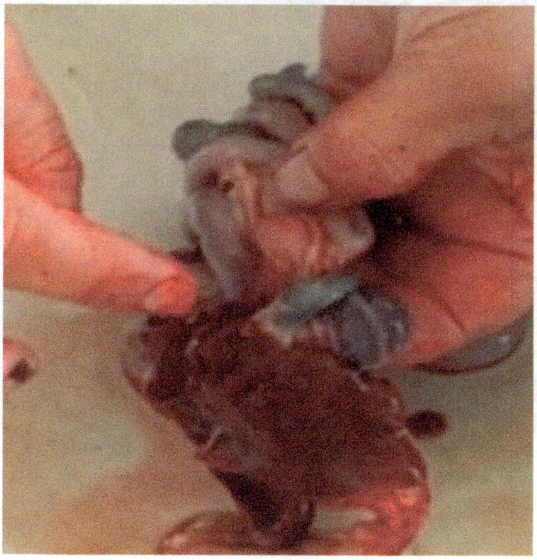

Figure 31 Gallbladder is bright green, do not puncture

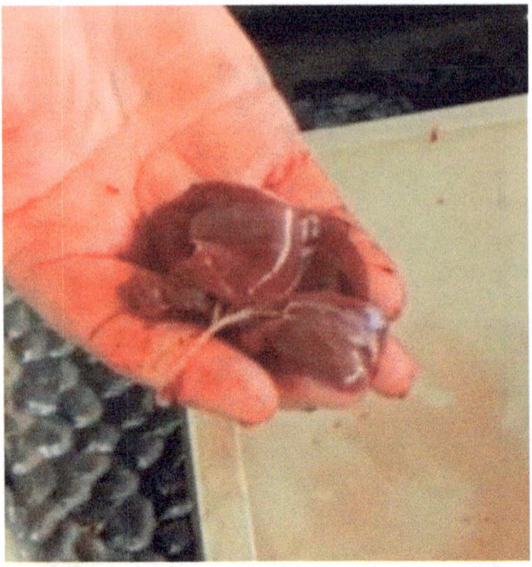

Figure 32 The liver

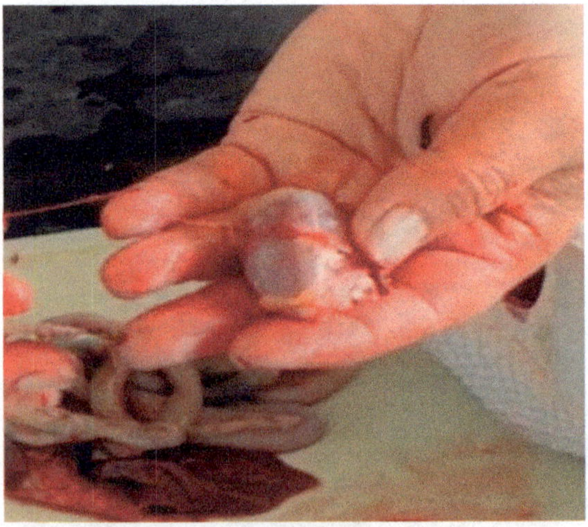

Figure 33 The heart

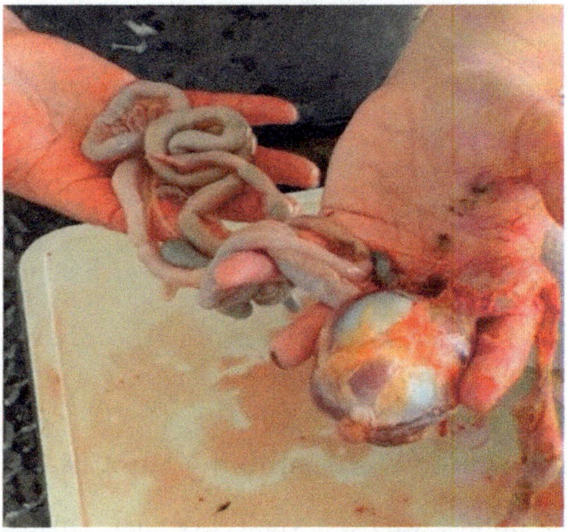

Figure 34 Stomach and Intestines

5. At this point, you may want to remove the neck. Depending on your plans for the bird, you can either keep it for making stock or as pet food. To remove the neck, simply cut it off at the base.

6. Rinse the carcass thoroughly at this point. You want to ensure that none of the innards are left on the carcass. Don't be shy about using a bit of water here. It is important that the carcass is clean. You should also be keeping your workspace clean in between each bird to prevent any contamination. Rinse your knives also; this will help prevent buildup that may be harder to remove once dry.

7. Once the carcass is clean, place the carcass in an ice bath in an esky or cooler. If you have a cool room, you can use that instead. You need to get the carcass temperature down to below 4°C (40°F) and let it rest at that temperature for at least 24 hours, preferably 48 hours. It is vital to ensure

the temperature stays below that. You can then freeze or refrigerate the carcass. Or go straight to cooking.

If you don't rest the carcass for at least 24 hours, we have found they can be a bit tough. This is because rigor mortis is still present. The carcass will relax between 24-48 hours, making for a very tender bird.

Step-by-Step Guide to Disassembling the Chicken

I don't break our birds down from whole, as even if I'm just after the meat, I prefer to cook the bird whole and then break the meat up.

For those that want to break birds down completely, here are the steps you need to follow. I would recommend finding a YouTube video to watch also, as it is not as easy as you think and watching an expert do it will be invaluable.

1. Remove the Legs: To remove the legs, cut through the joint where the leg meets the body. A good way to do this is by gently twisting the leg to pop the joint out of its socket. Once you've done that, you can remove the thigh and drumstick in one piece, or you can separate them if you prefer smaller cuts.

2. Remove the Wings: The wings are a bit easier to remove—just make a cut around the joint at the shoulder. You'll need to cut through some connective tissue to separate them from the body. You can leave the wings whole or break them down further into drumettes and wingettes, depending on your preferences.

3. The Breast: The breast is where the good stuff is. To remove the breasts, make a long cut along each side of the breastbone. Once you've cut through, carefully pull the meat away from the ribs. You may need to work your knife around the ribs to get all the meat off. You will need a sharp boning-style knife for this part.

4. Save the carcass: Don't throw away the carcass! It's great for making homemade chicken stock. Simply remove any leftover meat and save the bones and bits for a rich, flavourful broth.

Post-Gutting Cleanup

Cleaning up after the processing is done is a crucial step. It's not glamorous, but it's necessary to keep everything sanitary and safe for the next batch of meat you'll be processing. Here's what you need to do:

1. Clean Your Tools: After you've finished breaking down the chicken, make sure to wash your knives, cutting boards, and any other tools you used. You can use hot water and soap, followed by a disinfectant, to ensure everything is thoroughly sanitised.

2. Dispose of the Waste Properly: Compost any feathers, innards, and leftover bones if you can. If not, make sure to dispose of them in a proper waste disposal system. This will help reduce the risk of attracting pests and keep your area clean. I dig a big hole in my newest compost pile and bury them deep in there. Mark the area with a plant stake and ensure you don't turn it for at least 30 days, longer if possible. It will break down well, and there is a huge amount of nutrients in the offal and feathers.

Safe Handling and Storage

How to Clean and Store Your Meat

The final step in processing your chicken is storing it properly. The last thing you want is to ruin all your hard work by not handling the meat properly. Here's how to ensure your chicken stays fresh and safe for consumption.

1. Rinse and Dry the Chicken: After your bird is processed, rinse it one final time to ensure it's completely clean. Pat it dry with clean paper towels, as moisture can lead to bacterial growth and spoilage.

2. Storage in the Fridge: If you're planning to cook the chicken within a few days, you can store it in the refrigerator. Just make sure to keep it at the right temperature (below 4°C or 40°F) and keep it in a clean, airtight container to prevent any cross-contamination.

Freezing, Vacuum Sealing, and Preserving: Making Sure Your Hard Work Lasts

If you want to preserve your chicken for the long term, freezing is the way to go. But don't just throw the bird in a bag and call it a day—use a vacuum sealer to ensure the meat stays fresh and delicious. Vacuum sealing removes the air from the bags, which helps prevent freezer burn and ensures your meat lasts longer.

Vacuum Sealing: Once your chicken is dry, vacuum seal it in individual portions or whole. You can buy poultry shrink bags from Amazon. These are actually really simple to use. After placing the bird in the bag, wing end first, you dunk it in water that is heated to 90C (195F). The plastic will suck tight around the carcass to prevent freezer burn.

The bags come with straws that you placed into the cavity of the carcass while twisting the bag opening so you can push any excess air out of the bag. Once sealed, they are able to be frozen for about a year.

A couple of things to remember if using the shrink wrap bags. Ensure the pot is not filled so full that water will overflow. Also, make sure water doesn't get in the open end of the bag.

Don't forget to label each bag with the date so you know when it was processed. This will help you keep track of how long it's been in the freezer.

Storing in the Freezer: Place your vacuum-sealed chicken in the freezer, making sure it's stored at 0°F (-18°C) or lower. This will keep the meat fresh and ready for when you're ready to cook it.

Wrapping It Up

You did it! Processing your own chickens is one of the most rewarding parts of homesteading. It might be a little messy, and it definitely requires focus and patience, but when you sit down to enjoy that first meal, you'll know it was all worth it.

It will taste twice as good knowing that if you choose, you will never have to buy chickens from the store or supermarket ever again.

Chapter 6

Conclusion and Next Steps

Reflections on the Process

Well, here we are. If you've made it this far, I'm guessing you've processed your chickens—either you've got that fresh, home-raised meat ready for the table, or you're about to dive in and tackle the final steps. Either way, I want to take a minute to reflect on the journey you've been on.

I'll be honest—raising chickens for meat isn't a walk in the park. There are some tough moments, some messy moments, and a few surprises along the way. But guess what? You've made it! And that's something to be proud of.

Let's talk about what went well and what could have been better, because no one's perfect. The key is to keep learning, keep improving, and keep moving forward.

What Went Well

First off, let's take a moment to celebrate what went right. Raising chickens for meat can be hard work, especially if you are a working parent/partner with lots of time-consuming commitments. But I bet you did it with more confidence and skill than you probably thought you had at the start.

The fact that you've gone through the entire process—getting your chicks, setting up the brooder, transitioning them to pasture, feeding them for flavour, and then processing them—shows that you've taken on a significant challenge and followed through. That's huge, so take a moment to give yourself a high five. I think it's really important to acknowledge the effort you have made to feed your family healthy, homegrown meat.

I bet you learnt a lot along the way. Maybe you realised that chickens are smarter than you thought, or maybe you discovered that you're more capable of handling the dispatching process than you ever gave yourself credit for. If you followed all the steps and did your best, then you did it right. Even if things didn't go perfectly, you learnt something valuable from each stage of the process.

You likely learnt how to set up a proper brooder, how to handle your chicks with care, how to give them the best life possible, and how to process them in a humane and respectful way. These are all skills you'll carry with you for years to come, even if you decide that growing chickens is not what you love. And if that's the case, well, that is ok too!

Don't forget, this process isn't just about raising chickens—it's about gaining confidence in your ability to feed your family with your own hard work. That's a big win.

What Could Have Been Better

Now, let's talk about the things that didn't go quite as smoothly as you might've hoped. And let's be real here—no one's perfect, so there's no shame in admitting that some parts of the process didn't go exactly as planned.

Maybe you had a few hiccups with getting your chicks settled into the brooder. Perhaps the transition to pasture didn't go as smoothly as you'd expected, or maybe you found that processing the chickens took a bit longer than anticipated. That's okay. These things happen.

Maybe you struggled with the dispatching process. It's natural to feel a little uneasy about it, especially if it's your first time. But here's the thing: every time you go through it, you get better.

Next time, you'll know exactly how to set up your workstation more efficiently or perhaps how to be a little quicker and more confident. You're not going to get everything perfect on the first try—and that's okay. The goal is to learn from those moments and improve for the future.

If you ran into any setbacks, don't beat yourself up about it. Every single step in this process is an opportunity to learn and grow.

Think of it this way: If everything had gone perfectly, you wouldn't have learnt as much. It's the mistakes and the moments that make you question yourself that will ultimately make you a better, more capable homesteader.

Years ago, I bought my children a poster for their room; it says,

"Mistakes are proof you are trying."

So, take a deep breath, reflect on what could've been better, and then use that knowledge to make your next round of chicken-raising even smoother.

Encouragement to Keep Going: You've Raised Your Chickens; Now Enjoy the Rewards!

I want you to take a moment to congratulate yourself. You've just raised and processed your own chickens for meat. That's a tremendous achievement.

Not only have you gained new skills, but you've also taken a giant step toward self-sufficiency.

You've done something that most people will never experience, and now you get to enjoy the rewards of your hard work.

But here's the kicker: the work doesn't end here. Just because you've processed your chickens doesn't mean your journey is over. Far from it. This is just the beginning. You've proven to yourself that you have the ability to take control of your food supply and make the tough decisions that come with homesteading.

This experience will stay with you and will be the foundation for the next steps in your homesteading adventure.

Figure 35 One of our first homemade chickens

As you sit down to your first meal of home-raised chicken, I want you to savour the moment. Know that you've earned it. The taste will be richer, the satisfaction will be deeper, and the knowledge that you've provided this meal for your family will make it all the more special.

It's a feeling of true accomplishment, and it's one that only a few people ever get to experience. But guess what? You're one of them. You're a homesteader now. Own it!

Resources for Further Learning

Now, let's talk about the next steps. You've raised your chickens, processed them, and learnt a whole lot along the way. But the journey of homesteading doesn't stop here—far from it.

If anything, this is just the start of your path toward self-sufficiency and living the life you want. There's always more to learn, and as a homesteader, you're in the perfect position to keep growing, expanding, and refining your skills.

If you're looking to dive deeper into raising meat chickens, processing other livestock, or expanding your homesteading knowledge, here are a few resources that I love to help you along the way.

Think of them as your toolkit for the next phase of your adventure.

Whether you're looking for more hands-on advice, inspiration, or communities of like-minded folks, these resources will guide you through the next steps of your journey.

Books

Books are a fantastic way to deepen your knowledge, and obviously if you are reading this one, then you like to read. Here are a few I recommend that have helped me and many other homesteaders along the way:

- You Can Farm: The Entrepreneur's Guide to Start and Succeed in a Farm Enterprise by Joel Salatin

This was one of the first books I read when I started my homesteading journey. It's still relevant today, and we could probably refer to Joel as the Godfather of modern homesteading.

- Pastured Poultry Profit$ by Joel Salatin

The go-to book for those wishing to scale up their pastured poultry farm. I'm not even there yet, but it's my end goal. Great advice for those wishing to run a business.

Websites

If you're more of an online learner, there's no shortage of websites and blogs that can help you continue your journey toward self-sufficiency. These sites are packed with useful articles, videos, and even forums where you can ask questions and get advice from fellow homesteaders.

Mother Earth News: This site is a treasure trove of information on everything from gardening to raising livestock. The articles are well-researched, and the advice is practical and hands-on.

Polyface Farms: Run by Joel Salatin, a pioneer in the world of pasture-based livestock farming, Polyface Farms offers a wealth of information on raising animals in a sustainable way. Whether you're looking to learn about rotational grazing or how to build a mobile chicken tractor, this site has it all.

Mojo Homestead: Of course I am biased on this one. My website is slowly growing to include lots of information that budding homesteaders need. I talk about chickens, goats, bees, and cows mostly, but new stuff is added all the time. I also send out a weekly newsletter and a weekly podcast called Not The Farmers Wife.

Communities

The best part of homesteading is the community. There's a sense of camaraderie among people who are working hard to live more self-sufficient lives. Whether online or in person, getting involved with other homesteaders is a great way to continue learning and get support when you need it.

Facebook Groups: There are countless Facebook groups for homesteaders. Whether you're into raising chickens, beekeeping, or organic gardening, you can find a group for just about anything. I recommend searching for a few local homesteading groups and joining the conversation. Don't forget to visit the Mojo Homestead Facebook page. Be sure to post pictures of your chicken processing and tag me in!

Let's Get More Self-Sufficient Together!

By now, you've proven to yourself that you can take care of your chickens from start to finish. You've done the work, faced the challenges, and come out stronger on the other side. But this is only the beginning.

There's so much more to explore in the world of self-sufficiency and homesteading. Whether you want to raise more animals, grow your own food, or just become more self-reliant in your daily life, the possibilities are endless.

Don't stop now. Take what you've learnt, build on it, and keep moving forward.

You've got this.

And remember, there's an entire community of homesteaders out there cheering you on every step of the way. We're all in this together, and we're only stronger when we share our knowledge, experiences, and encouragement.

So, what's next? The sky's the limit. Go ahead and tackle the next step of your self-sufficiency journey, knowing that you have the skills, the drive, and the passion to succeed.

Whether it's raising more livestock, expanding your garden, or diving into another aspect of homesteading, I'm excited for what's to come.

Keep learning, keep growing, and keep enjoying the rewards of the life you're building.

Thank you for letting me into your life to help you on this self-sufficiency journey. Let's get more self-sufficient together!

What Next?

Now that you have mastered raising your own meat birds, what's next?

The next book in the series is coming soon. Titled "Eggs without Cartons", I think you can guess that is about the path to take to break up with supermarket eggs. It will be a guide to growing eggs in your backyard.

Because life is too short for store bought eggs.

Made in the USA
Monee, IL
03 November 2026

67385894R00056